U.S Environmental Protection Agency

Cleaner Diesels: Low Cost Ways
to Reduce Emissions from Construction Equipment

March 2007

The information contained in this report
was prepared as part of EPA Contract EP-W-5-022 and EPA Contract 68-W-03-028

Prepared for:
U.S. Environmental Protection Agency
National Center for Environmental Innovation

Prepared by:
ICF International
9300 Lee Highway
Fairfax, VA 22031
(703) 934-3000

Table of Contents

1 Introduction

1.1 Background

Air pollution from diesel emissions is a public health concern that reaches every part of the country. There are two main pollutants of concern in diesel exhaust that affect human health: nitrogen oxide (NOx) and particulate matter (PM).

- NOx is one of the main ingredients in the formation of ground-level ozone, which can trigger respiratory problems. Ozone can aggravate asthma and other respiratory diseases, leading to more visits to the emergency room and increased hospitalizations. Ozone can inflame and damage the lining of the lungs. This may lead to permanent changes in lung tissue and to irreversible reductions in lung function if the inflammation occurs repeatedly over a long time period.

- PM has been associated with an increased risk of premature mortality, hospital admissions for heart and lung disease, and increased respiratory symptoms. Long-term exposure to diesel exhaust is likely to pose a lung cancer hazard. In addition, PM, NOx, and ozone adversely affect the environment in various ways including visibility impairment, crop damage, and acid rain.

The construction sector is a significant contributor to these emissions, creating 32 percent of all mobile source NOx emissions and 37 percent of PM emissions.[1] While stringent new emissions standards are scheduled to significantly reduce emissions from new nonroad equipment starting in 2008, much of the equipment in the current nonroad diesel fleet will continue to operate for many years to come. Therefore, reducing emissions from the existing legacy construction equipment fleet is an important component of EPA's emissions control strategy.

The construction sector is highly diverse and is made up predominately of smaller companies. Approximately 92 percent of construction companies have 20 or fewer employees. They tend to be low-margin businesses, with much of their business value accumulated in their capital equipment. Consequently, construction companies resist modifications that they believe will restrict their equipment's operability or increase maintenance. Small companies may not have the ability to spend significant resources to reduce emissions from their equipment.

1.2 Purpose

The purpose of this research project was to study and identify low cost ways to reduce emissions from nonroad construction equipment. The report documents the costs and benefits of a number of these strategies – actions that may be taken by small companies (and medium or larger ones as well) in the construction sector to reduce their emissions.

---

[1] Recommendations for Reducing Emissions from the Legacy Diesel Fleet. October 7, 2005, http://www.epa.gov/cleandiesel/documents/caaac-apr06.pdf.

Through our research we found there are a variety of operating practices and technologies that companies can employ at low cost. In many cases, strategies such as reduced idling or better preventive maintenance can help lower operating costs while also reducing emissions. Companies that voluntarily participate in programs to improve the environment and reduce air emissions benefit from an improved public image and better community relations. Reductions in diesel exhaust at construction sites can lower the incidence of respiratory problems in surrounding communities, improve the workplace environment, and contribute to improvements in regional haze and other environmental impacts associated with emissions from diesel engines.

1.3 Research approach

Primary research was conducted through telephone contacts with key industry associations, engine and equipment manufacturers, technology experts, and government and business officials. We also obtained information from a review of secondary sources, including trade publications, government reports, manufacturer web sites, and other publicly available sources.

We sought to identify both the business and environmental benefits of the strategies studied. The construction sector contains a diverse array of equipment types, specialized companies, and operating practices. In many cases, our research uncovered cost or benefit information that was specific to particular types of companies or construction market niches. This report reflects that diversity and provides appropriate context and caveats for the information. Detailed quantitative studies were often unavailable; in these cases it was necessary to rely on anecdotal information or to extrapolate from related research.

1.4 Report outline

The report groups low cost activities in three categories: (1) operating strategies, (2) fuel strategies, and (3) equipment strategies. Operating strategies in Section 2 include reducing unnecessary idling, improving preventive maintenance, and training equipment operators. Section 3 focuses on use of cleaner fuels, including ultra-low sulfur diesel and biodiesel. Equipment strategies in Section 4 include retrofits, repowering/engine upgrades, and electrification. For each strategy, we provide a brief description, report cost/benefit information, and discuss practical implementation issues. Section 5 summarizes our conclusions.

2 Operating Strategies

This section describes three operating strategies to reduce diesel emissions: (1) equipment idle control and reduction, (2) engine preventive maintenance, and (3) equipment operator training. Each offers contractors a way to reduce diesel emissions while also achieving significant reductions in operating costs that will improve their bottom line. Companies can implement all three strategies simultaneously.

The table below summarizes the costs and benefits of each operating strategy. Sections 2.1 through 2.3 provide more detailed information on each of the three strategies, including their costs, benefits, and how to do it.

Operating Strategies Summary

Operating Strategy	Costs	Benefits
Equipment Idle Reduction and Control	Low administrative costs for training and tracking of idling If on-board idle reduction equipment is used, upfront investment in equipment is required	Reduced PM, NOx, carbon monoxide (CO), and HC emissions Significant fuel cost savings Longer engine life and reduced maintenance costs
Engine Preventive Maintenance	Low administrative costs for tracking equipment maintenance needs If customized software is used to track maintenance, significant upfront investment in software may be required	Reduced PM, NOx, CO, and HC emissions Reduced fuel consumption Reduction in high cost engine failures Longer equipment life and reduced maintenance costs
Equipment Operator Training	Upfront investment in operator training – cost varies by training program	Reduced PM, NOx, CO, and HC emissions Improved operator efficiency Reduced fuel consumption

2.1 Equipment Idle Reduction and Control

Elimination of unnecessary idling can save fuel, prolong engine life, and reduce emissions. It can also help reduce the noise levels associated with construction. Unnecessary idling occurs when trucks wait for extended periods of time to load or unload materials or supplies, or when equipment is left on when it is not being used. Workers may take breaks and leave equipment running unnecessarily or may idle equipment because it is an ingrained habit. Many workers may be unaware that most pieces of construction equipment do not require extended warm-up and cool-downs. In other situations, workers may unnecessarily idle equipment because they are unaware of the cost impact of this practice on the company and its equipment. Managing equipment operations and training workers to reduce unnecessary idling is a relatively easy way to lower operating costs and help reduce the environmental impact of construction. These actions result in cleaner air and health benefits for workers at the site.

2.1.1 Costs

The cost of reducing idling varies according to the strategy employed. A contractor can implement a company idling policy as a low cost solution. This can involve simply raising awareness among equipment operators and managers of how much unnecessary idling is costing the company and advising operators to turn off equipment that is not being used. Costs are low for this type of program, and the level of effort can be tailored toward the opportunity companies see for cost savings.

"The engines in many construction trucks do more than just move the truck. They also spin concrete drums, pump off cement, lift pallets of bricks, bags and other supplies, run diggers and man buckets, and perform a myriad of other tasks. So their engines don't often shut off....But the truth is, drivers go too fast, idle engines many times when they are doing no work, rev them higher than they should, and in general blow any possible fuel savings out the stack. They do so because they have never been told not to or because they think it's what their engines need, or it's what they and their buddies do." Tom Berg, Editor, Construction Equipment.[2]

Some nonroad equipment is idled to run cab accessories, such as heating and air conditioning. While the use of auxiliary power units (APU) is more common in onroad trucks, manufacturers have begun to market this equipment to nonroad equipment users as well. There are limited opportunities to employ APUs in the construction sector, but companies may have some equipment on which they can be used. Equipment operating in extreme conditions where the vehicle is idled extensively to maintain cab comfort would be a target application for this technology. Caterpillar's MorElectric system can be installed in both onroad and nonroad equipment.[3] A variety of products are marketed by different vendors. The cost of the equipment ranges from $500 to $9,000. A list of some idle reduction technologies for the trucking sector (direct fired heaters and auxiliary power units) and relevant cost information is provided at:

http://www.epa.gov/smartway/idlingtechnologies.htm#truck-mobile

---

[2] Berg, Tom. "How to Stop Idle Waste of Fuel." Construction Equipment. Oct. 2004. Vol. 107, Iss. 10.
[3] http://www.cat.com/cda/components/fullArticleNoNav?ids=202795&languageId=7.

In onroad vehicles, the greatest savings from idle reduction equipment come from reducing the need to idle to maintain cab comfort when the driver is sleeping. Idle reduction in nonroad equipment typically is achieved by reducing unnecessary idling that occurs during the work day. The Argonne National Laboratory has developed a fuel savings calculator that allows companies to estimate the costs and benefits of purchasing idle reduction technology. Although the worksheet is tailored toward onroad vehicles, it has information relevant to nonroad vehicles as well. The worksheet can be accessed at:

http://www.transportation.anl.gov/pdfs/EE/361.pdf

2.1.2 Benefits

A typical idling diesel engine in an onroad tractor consumes 1.2 gallons of fuel per hour at high idle and 0.6 gallons per hour at low idle.[4] There is a lack of detailed data concerning nonroad equipment idling. Fuel consumption for nonroad equipment at idle varies by equipment type. A typical mid-size track-type tractor consumes approximately one gallon per hour at idle.[5] At current diesel prices, a vehicle with just a single hour of unnecessary idle time per day is wasting $360 - $720 of fuel per year.[6] A fleet with 50 pieces of equipment that reduces unnecessary idling by one hour for each piece of equipment would save $72 -$144 per day in fuel. Over the course of a 250 business day year, this could save a company $18,000 – $36,000 in fuel costs. Ken Katch, Director of Emissions Solutions Group at Caterpillar, notes, "The amount of time equipment spends idling on a jobsite can be used as one measure of productivity. So there are other benefits to examining idling time besides fuel savings and reduced emissions. Equipment owners should examine their idling practices to see if they are based on today's modern diesel technology or whether they are legacy practices that are costing them money."[7]

Grace Pacific in Hawaii has implemented a program to reduce unnecessary idling. Grace Pacific has compiled an inventory of their fuel use, idling time, and air emissions. The inventory provides a baseline for tracking performance of the company's diesel emissions reduction program. They believe they can cut their overall fuel consumption by 10 percent on Oahu, saving the company approximately $80,000 in fuel costs and reducing emissions substantially.[8]

For an onroad truck, eliminating one hour of idling reduces PM emissions by two grams, NOx emissions by 136 grams and CO_2 emissions by 6,848 grams.[9] For nonroad equipment, emissions benefits vary by equipment type. For a typical backhoe loader, reducing a single hour of

---

[4] U.S. EPA. Study of Exhaust Emissions from Idling Heavy-Duty Diesel Trucks and Commercially Available Idle-Reducing Devices. October 2002. http://www.epa.gov/otaq/smartway/documents/epaidlingtesting.pdf.

[5] Phone conversation with Ken Katch, Caterpillar, September 27, 2006.

[6] We assume the vehicle is operating at low idle and pays $2.40 a gallon for offroad diesel. The cost range incorporates the difference between low and high idle.

[7] Email communication, Ken Katch, Director of Emissions Solutions Group, Caterpillar, Inc, January 29, 2007.

[8] Email and phone contact with Chris Steele at Grace Pacific.

[9] http://www.epa.gov/oms/smartway/documents/epaidlingtesting.pdf.

unnecessary idling would reduce PM emissions by 13 grams, NOx emissions by 155 grams, CO emissions by 65 grams, and CO_2 emissions by a similar amount.[10]

Reducing idling will prolong equipment life. An idling engine does not generate enough heat to achieve proper combustion. Deposits can build up on the piston and cylinder walls, contaminating the oil and creating friction that wears out engine components faster. Diesel engines achieve optimum performance at a reasonably high RPM under load.

Many companies already have equipment that enables them to use either mechanical or electronic controls to automatically shut off engines when idling for more than a few minutes. Often companies only need to turn on these features. Companies should check with their equipment manufacturer to determine if these features are available in their products. Caterpillar has estimated that a midsized wheel loader idling approximately 30 percent of the day could reduce annual fuel costs by $656 by using its new idle management features.[11]

Limiting idling also lowers costs by reducing the need for maintenance. By reducing wear on the engine, idling less will reduce the need for oil changes and engine rebuilds, thereby lowering operating costs. Less idling also reduces employee and public exposure to unhealthy emissions, which can have a positive effect on employee health and productivity.

2.1.3 How to do it

The first step in reducing unnecessary idling is to define a policy and inform employees. Operators simply need to turn off equipment when it is not in use. According to Bob Lanham, Vice President, Williams Brothers Construction, "We approach our idle reduction policy from a behavioral standpoint. If you get off of the equipment, you turn it off. You enforce that through supervision. If you are not burning fuel, you are saving money. It's good for the environment. From a safety standpoint there is no chance for a piece of equipment to accidentally engage."[12]

An idle reduction policy can also include measures to mitigate exposure to idling equipment. For instance, a staging area for vehicles waiting to access the site can be set up away from high volume pedestrian areas or other public spaces. To the extent possible, generators and other equipment should be located away from fresh air intakes on occupied buildings.

Operator training is an important part of any idle reduction plan. Operators need to understand the needs of their equipment, how they can reduce idling, and how it will serve the goals of the company. Bob Lanham of Williams Brothers Construction notes, "We promote idle reduction in three different ways…the environment, safety and cost. When we make the appeal that way, we

---

[10] We assume an uncontrolled backhoe loader, with an 89 horsepower engine, operating under a load factor of 0.21. Data obtained from EPA report, Exhaust and Crankcase Emission Factors for Nonroad Engine Modeling—Compression-Ignition. Report No. NR-009c, Revised April 2004.

[11] New Engine Idle Management System Offers Increased Fuel Efficiency With Customer Flexibility for Caterpillar Midsize Wheel Loaders, http://www.cat.com/cda/components/fullArticleNoNav?ids=209563&languageId=7.

[12] Phone conversation, Bob Lanham, Vice President, Williams Brothers Construction Co., Inc, January 19, 2007.

can paint a picture that influences the greatest number of people, depending on what their values are, and the message has the greatest chance to modify behavior."[13]

The idle reduction plan should define required warm-up and cool-down periods for equipment. Check the owners' manuals or contact your equipment manufacturer to determine the appropriate warm-up and cool-down periods. For the trucking sector, older engines will require a three to five minute engine cool down. Newer equipment requires almost none. Morning warm-up periods should also be restricted to three to five minutes.[14] Dump trucks and supply/delivery vehicles waiting to load or unload for greater than five minutes can be shut down.

Some equipment has idle management systems built in. For instance, Caterpillar's new Engine Idle Management System for its H-Series Wheel Loaders has four different control settings to manage idle time. The work setting allows operators to adjust idle speeds between 650 RPM and 1000 RPM. The "hibernate" mode is engaged when the transmission is in neutral, the parking brake is set, and the fan current is greater than 0.8 amps. The "warm-up" mode is used to keep the machine warm in cold weather. It increases idle speeds in cold weather based on electronic monitoring of coolant and the inlet manifold temperatures. A "low voltage" mode increases engine speed when the battery drops below a specific voltage threshold.

Many pieces of equipment come with automatic shutdown features. These allow for the automatic shutdown of vehicles after a fixed period of time. Electronic controls can be programmed to automatically shut down the engine once it has been operating at a specified RPM for a preset amount of time. The controls may include a programmable load factor that prevents an idling machine from shutting down if it is operating an attached device. Often companies program shutdown if the clutch, brake, and accelerator pedal are not touched for five minutes. Appropriate shutdown specifications can be applied to different equipment types. Contractors should determine if their equipment has mechanical or electronic controls allowing for automatic shutdown and enable these features where appropriate.

Another important component of an idle reduction policy is measuring performance. While there are administrative costs to tracking fuel consumption by equipment operator, some companies use software that can be set up to collect this information. Many companies like to post results so that operators are aware of how they compare to others.

An idle management policy can also seek to better manage vehicles and equipment that are accessing the construction site. If significant idling is occurring while vehicles are delivering supplies, better scheduling of pickups or deliveries could help alleviate such idling. "Idle reduction opportunities will vary by equipment type and operation. You should pay close attention to the job production cycle. Significant idle reduction can be achieved by more efficiently managing the flow of work within a project."[15]

---

[13] Phone conversation, Bob Lanham, Vice President, Williams Brothers Construction Co., Inc., January 19, 2007.
[14] EPA New England. What You Should Know about Truck Engine Idling. April 2002. http://epa.gov/NE/eco/diesel/assets/pdfs/Diesel_Factsheet_Truck_Idling.pdf.
[15] Phone conversation, Terry Goff, Director Public Policy & Regulatory Affairs, Caterpillar, January 19, 2007.

2.2 Engine Preventive Maintenance

A preventive maintenance program seeks to maintain engines at their original level of performance and eliminate the high cost of catastrophic engine failure. Preventive maintenance is the systematic inspection, detection, and correction of potential equipment failures. It includes many different elements, such as an inventory of equipment, corporate policies to implement periodic equipment maintenance, and training for operators and mechanics so they can detect problems early.

An effective program should include a plan for managing each piece of equipment over its lifetime. This requires an inventory of the periodic maintenance requirements for each piece of equipment and accurate measurements of the hours of use. Based on equipment usage tracking and maintenance requirements, companies can appropriately schedule preventive maintenance. Simple maintenance to improve equipment efficiency and engine life includes air/fuel/oil filter replacement, battery replacement before failure, and regular oil changes.

2.2.1 Costs

There are a number of different approaches to implementing preventive maintenance programs. Small contractors have stressed that good management and record keeping can accomplish the goals of preventive maintenance. The only administrative cost for these simple programs is labor time to track maintenance requirements. Using spreadsheets to keep track of equipment maintenance data and stickers on equipment to record the last and next service required has worked well for some small companies, and at very low cost.[17]

Some large companies use custom software solutions to ensure consistency in the implementation of preventive maintenance programs. Prices for software vary by vendor, the features purchased, and the number of licenses purchased. One company with 96 mechanics estimated that fleet management software for this size of operation can range between $100,000-$150,000. Another company reported annual licensing costs of $5,000 for a staff of 30 managing 1,400 vehicles. Initial setup costs were estimated at $80,000-$100,000.[18]

"It's surprising, the cost of shipping a broken down machine…a machine that comes to a remote site with dead batteries can take a day and a half to get running. If that's a key piece on a job with $40,000 per day liquidated damages, you just spent $60,000 replacing a battery…even if it's just a run-of-the mill loader, you have an operator and an oiler standing around waiting while your mechanic looks the machine over. The project super has to go rent a replacement….This thing's got long tentacles. When you multiply all those man hours by a $55-per hour shop rate, you're talking about some real money." Thad Pirtle, Traylor Bros[16]

---

[16] Stewart, Larry. "Reliability Enlists Project Support for Maintenance." Construction Equipment. October 2004. Vol 107, Iss. 10, p. 59.

[17] Brown, Daniel. Preventive Maintenance Pays Dividends: How Six Contractors Handle their Equipment Maintenance. Concrete Construction. March 1, 2005.

[18] Bordenaro, Mike. "Fleet Management Software Evolves." Construction Equipment. May 2006. Vol. 109, Iss. 5, p. 50.

2.2.2 Benefits

The purpose of preventive maintenance is to maximize equipment life and minimize costly equipment failures. A recent study found that 38 percent of the overhead cost of operating construction equipment was attributable to component failures (major failures would include bearings, rods, gears, etc).[19] A systematic maintenance program can prevent the performance of unnecessary or premature maintenance as well as the need for repairs after catastrophic failures. Greg Sitek of Reed Construction Data notes, "You should have a maintenance plan or program before you buy your first piece of equipment. The return on your equipment investment is tied directly to how well it is cared for."[20]

Proper maintenance also significantly reduces fuel consumption and emissions. Likely fuel savings vary across equipment types. Basic maintenance, such has changing the oil and oil filter at proper intervals, can save fuel through maintaining the lubricating properties of oil. Fuel economy improvements of two to three percent due to improved oil filters have been recorded in highway tests.[21] Over-extended oil changes can also cause power losses, which translate into fuel economy losses. Power losses of 18 percent due to overextended oil changes have been shown in tests of Cummins engines.[22]

Some contractors have implemented software tracking and scheduling of preventive maintenance. Contractors have reported cutting the need for engine rebuilds in half following improvements in the management of preventive maintenance.[23] One company reported the use of equipment tracking software allowed them to more efficiently schedule required oil changes and other maintenance, reducing maintenance costs by about 15 percent.[24]

Companies that use oil analysis to improve preventive maintenance have reported savings in the form of reduced equipment repairs. One company reported sending about 1,000 samples of oil to a vendor in the course of the year at a cost of about $10,000. About 4 percent of these samples came back with a critical flaw requiring action. In one case, the oil analysis flagged a loader where a gear had come loose. Repairing the machine before a failure saved the company over $30,000 in direct maintenance costs for this single machine.[25] Another company, Kimmins Contracting, reported savings of $300,000 though the use of oil sample analysis. They were able

---

[19] Waggoner, Stephen. "Boost Utilization Rates with Effective Oil Management." Cranes Today. March 2006.

[20] Sitek, Greg. "Equipment Maintenance." Reed Construction Data, January 16, 2006.

[21] Fitch, Jim. "Clean Oil Reduces Engine Fuel Consumption." Maintenance World. December 13, 2004. http://www.maintenanceworld.com/Articles/noria/clean-oil-fuel-consumption-poa2.htm.

[22] Fitch, Jim. "Clean Oil Reduces Engine Fuel Consumption." Maintenance World. December 13, 2004. http://www.maintenanceworld.com/Articles/noria/clean-oil-fuel-consumption-poa2.htm.

[23] Stewart, Larry. "Barber Brothers reforms maintenance with tools in hand." Construction Equipment. August 2003. Vol. 106, Iss. 8, p. 71, http://www.constructionequipment.com/.

[24] Stewart, Larry. "Fleet Software Drops Labor Cost 15 Percent." Construction Equipment. May 2005. http://www.constructionequipment.com/.

[25] Brown, Daniel. Preventive Maintenance Pays Dividends: How Six Contractors Handle their Equipment Maintenance. Concrete Construction. March 1, 2005. http://www.concreteconstructiononline.com/industry-news.asp?sectionID=718&articleID=239491.

to reduce catastrophic equipment failures like broken or bent rods through preemptively detecting problems.[26]

There are significant emissions impacts from improperly maintained diesel engines. Limited information is available for nonroad equipment, but two recent studies of onroad equipment shed some light on the importance of proper maintenance in diesel engines. An EPA study of onroad heavy diesel engines shows improperly maintained equipment can cause increases in CO, NOx and PM emissions. The EPA study simulated a number of different engine problems that might be experienced due to a lack of preventive maintenance.[27] Nozzle hole wear in fuel injectors increased CO emissions by 40 percent and PM emissions by up to 85 percent. The study also simulated a loss of intercooler efficiency due to plugging and fouling. Intercooler fouling caused NOx emissions to increase by 7 percent and CO emissions to increase by 10 percent. Increased lube oil consumption was shown to increase PM emissions by approximately 85 percent, while also marginally increasing emissions of HC and CO. While these emissions test results are most applicable to onroad diesel trucks, they do indicate that preventive maintenance could significantly decrease emissions from all diesel equipment.[28]

Another study, conducted for the California Air Resources Board (CARB), also estimated that poor maintenance can substantially increase emissions in onroad heavy-duty diesel trucks. Clogged air filters can increase PM emissions by 40-50 percent. Minor injector problems can increase PM emissions by 35-75 percent. Excess oil consumption can increase PM emissions over 100 percent. While these estimates were not calculated for nonroad equipment, they indicate the magnitude of emissions benefits that basic maintenance can have for large diesel engines.[29] A general conclusion is that higher emissions and oil consumption typically translate to lower efficiency and increased fuel consumption.

2.2.3 How to do it

All equipment owners can implement the basic elements of a preventive maintenance program. Creating a database/inventory of equipment and periodic maintenance requirements is the starting point for such a program. Accurately tracking equipment usage is also important. This can be done manually, but some companies have found it profitable to implement automatic vehicle tracking systems to allow equipment location and usage to be recorded electronically.

Many companies use software to manage the preventive maintenance process. Some large fleets use equipment modules in enterprise tracking software.[30] Other companies use fleet management software purchased from vendors or software they designed in-house. Such software flags

---

[26] Stewart, Larry, "Kimmins Saves $300,000 with Oil Analysis." Construction Equipment. November 1, 2005. http://www.constructionequipment.com/.

[27] The test engine selected was a 2000 model year Cummins ISM 350 ESP. Two different heavy-duty transient cycles were employed to simulate typical operation.

[28] Sharp, Christopher. "Transient Emissions Characterization of Simulated Diesel Engine and Component Failures." EPA, September 2001.

[29] Heavy Truck Emissions Factors Development. California Air Resources Board, May 15, 2002. http://www.arb.ca.gov/msei/onroad/downloads/tsd/HDT_Emissions_New.pdf.

[30] Enterprise Software is software that solves an enterprise problem (rather than a departmental problem) and usually enterprise software is written using Enterprise Software Architecture.

equipment for 250-hour maintenance intervals and provides reports of maintenance that needs to be performed each week. Fuel monitoring systems can be integrated with some fleet management software packages. Some companies have set up severity-based maintenance systems that trigger preventive maintenance based on fuel usage calculations. Monitoring fuel consumption and oil consumption can help identify problems. According to Ronnie Falgut of Barber Brothers Contracting, "When we went out into the field, we were finding filters on machines that had year old dates on them…we were trying to track over 200 pieces of equipment by hand, and that's just too much for any one person to take care of without some kind of record-keeping system."[31]

Smaller contractors can improve preventive maintenance merely through better record keeping or use of a spreadsheet to track the maintenance requirements. Companies should make sure they are tracking all of the information they need to make informed maintenance decisions. Records should include the make and model of equipment, the date and miles/hours at the time of the last service, and the details of service completed to specific components. Engine manufactuers recommend that fleet owners include preventive maintenance practices for each piece of equipment on the spreadsheet.

When companies track incidents of unscheduled maintenance, they can identify trends in the data. These trends might include determining if machines are susceptible to problems on certain components or using the data to develop estimates of service life for different components. This information can then be used to adjust preventive maintenance programs as needed.[33]

Basic preventive maintenance also requires companies to institute policies and procedures to identify the signs of equipment failure before they occur. Building a company culture where operators take pride in the maintenance and upkeep of their equipment is important. Operators can have a large impact on maintenance costs by being vigilant in identifying abnormal equipment operations. As noted by Dave Terres of Cold Spring Granite, "We had operators who were afraid to say anything about equipment problems…now we're telling them that if they hear or see or feel something that seems wrong, they should bring it to our attention right away."[34]

> "Operators should learn to trust their instincts and senses about what's happening with the machines…the smell of hot wiring means you more than likely have a short, and you'll want to fix it before it melts part of the paver's wiring…similarly, vibrations you feel during operation might be coming from a bearing point going out. It's a lot easier to replace a bearing right away than have it fail and have to replace it and other related components it may have damaged." Brodie Hutchins, Ingersoll Rand[32]

[31] Stewart, Larry. "Barber Brothers Reforms Maintenance with Tools in Hand." Construction Equipment. August, 2003. Vol. 106, Iss. 8, p. 71.

[32] "Help Your Paver Live Its Full Life." The Asphalt Contractor. March 2006. Vol. 20, Iss. 3, p. 32.

[33] Schultz, Becky. "Polish up your PM Program." Equipment Today. May 2006. Vol. 42, Iss. 5, p. 6.

[34] Stewart, Larry. "Maintenance Reduces Fleet Size." Construction Equipment. September 2003. Vol. 106, Iss. 9, p. 68.

The incidence of equipment failures can be reduced by implementing a policy that seeks to fix smoking equipment before it fails. Williams Brothers Construction in Houston has a "no smoking" policy, under which operators are trained to identify equipment that is producing abnormal smoke in the exhaust. Smoking equipment is flagged for further inspection and possible service. Typically, blue exhaust gas will indicate oil consumption under a low load operation. Black smoke is related to over-fueling, when the engine is operating at full load and a high temperature.[35] Black smoke may indicate that engine maintenance is required.

It is also important to train operators to inspect their vehicles daily for tire pressure, fluid leaks, fluid levels (engine oil, coolant level, transmission fluid), oil color, or other elements recommended in the owner's manual. Companies should work with their dealer or distributor to develop daily check lists for their drivers.

Many companies use oil and coolant sample analyses to identify equipment that may require overhauls or tuning. Oil conditions such as oxidation, additive depletion, and viscosity changes can be detected. These are usually caused by engine overheating or overextension of the oil change interval. Moisture or dirt contamination can also be detected. Mechanical problems, such as plugged air filters, blocked fuel lines, weak injector springs, and dirty injector tips, can be detected with oil sample analysis. Operational problems like lugging and over speeding can be uncovered.

---

[35] Over-fueling occurs when the fuel-air ratio is too rich, causing incomplete combustion of fuel in the cylinder.

2.3 Equipment Operator Training

Many companies train their equipment operators to enhance their skills. Operator training can provide a range of business benefits, while also reducing fuel consumption and emissions. Equipment training typically addresses a broad range of issues, including operating equipment in a safe and efficient manner, maximizing the productive capacity of equipment to do work, and being knowledgeable of the capability and limits of equipment. Some companies set up their own in-house training programs, while others choose to purchase training services from equipment manufacturers, equipment dealers, or other third parties. The amount of course material directly related to reducing fuel consumption varies.

2.3.1 Costs

The cost of training courses varies by manufacturer and the entity that is providing training. Caterpillar offers an extensive set of training programs for operators. The cost for a typical course is approximately $1,500 per person for a two-and-a-half day course. The course provides classroom and hands on training, simulators, Machine Application and Performance Seminars (MAPS), and certification. Training for a variety of equipment types is offered, including track-type tractors, wheel loaders, front shovels, hydraulic excavators, wheel tractor scrapers, backhoe loaders, articulated and off-highway trucks, and Challenger tractors. Training classes for motor graders last five days and cost $3,000. Non-certified courses are offered for $600 per day per person. Training is conducted in Peoria or at a dealer site if companies are willing to pay for travel costs for the trainers.[36]

Bobcat Co.'s training program uses operator training kits. Kits are available for excavator, skid-steer loader, VersaHandler, backhoe and planer attachment, safety and service safety training. The courses can be administered by anyone, including the dealer if desired. The kits range in cost from $33 to $150 and take approximately four hours to complete.

VISTA Training's TIPS from the Pros videotape series helps experienced operators refine their skills. The cost is $150. Training materials are available for a number of different equipment types.[37] Simple changes in equipment operation can increase productivity dramatically.

2.3.2 Benefits

Effective operator training increases productivity, provides for a safe work environment, reduces maintenance costs, and lowers machine fuel consumption. George Schulz, a Certified Dealer Instructor for Giles & Ransome notes, "Certified Equipment Training now allows buyers to make an investment in their operators that will pay huge dividends. The cost of the training is miniscule when compared to replacing an undercarriage on a large dozer or fixing a blown tire

---

[36] Ostrowski, Christopher. "Equipment Training Programs Vary by Industry, Manufacturer." Texas Construction. April 20, 2002. Vol. 10, Iss. 4, p. 49.
[37] Equipment Productivity Techniques. VISTA Training Programs for the Construction and Surface Mining Industries. http://www.vista-start-smart.com/html/tips_from_the_pros.html.

on an off-highway truck."[38] Enhanced efficiency allows jobs to be executed in a more timely fashion and reduces the amount of time that equipment is operated, thereby reducing air emissions.

Training can pay off by teaching employees to operate their equipment in a manner that minimizes the amount of time it takes to do a job. Employees learn how to operate their equipment close to the "sweet spot" where engine performance is optimized. Even experienced operators can improve their productivity by five percent. Improvements for less experienced operators can be even greater. [39] George Schulz of Giles & Ransome notes, "When operators learn how to utilize their machines properly, they will increase production and help to complete jobs ahead of schedule. Using controls in the operator station properly will make operators more efficient and lower unit costs."[40]

One example of productivity and profit improvement is the experience of a contractor who estimated that excavator cost for a 4,985-foot pipe-laying job would be $2.26 per foot. This estimate assumed a 76,000-pound machine would dig a foot of trench every minute and 41 seconds. A minor improvement in digging methods cut nearly a day from the project and increased gross profit for the job by 33 percent.[41]

Since most operator training programs focus on both safe and efficient operation of equipment, another benefit of training and certifying operators is fewer accidents and reduced insurance rates. Insurance rates are often more favorable for businesses that require employees to complete equipment training. Trained operators may enable companies to more easily obtain new work. Businesses competing for a job are sometimes required to show proof of equipment training during the bidding process.

> "One point we make when we train operators is that all of the profit for a day on most jobs is made in half an hour...if they are unproductive for half an hour, the job loses money for the day, but if they can find a way to improve productivity just a little, they can easily double the profit." Rich Deeds, Brubacher Excavating[42]

Improved operational efficiencies can reduce emissions and save money through reduced fuel consumption. Operators who are able to finish jobs five percent faster are achieving similar percentage emissions reductions and fuel savings by reducing the amount of time equipment is being used. Over the course of a year, a five percent increase in operator efficiency for a backhoe loader could save a firm $375 in fuel costs.[43] Equipment manufacturers believe that operating techniques for tractors, such as slot dozing (discussed in more detail below), could increase machine productivity by as much as 20 percent.

---

[38] "Giles & Ransome Touts CAT Certified Operator Training." ConstructionEquipmentGuide.com. 6/4/2003.

[39] Stewart, Larry. "Production Heroes: Take the Textbook to the Trench." Construction Equipment. April 23, 2003. Vol. 106, Iss. 4.

[40] "Giles & Ransome Touts CAT Certified Operator Training." ConstructionEquipmentGuide.com. 6/4/2003.

[41] Stewart, Larry. "Production Heroes: Take the Textbook to the Trench." Construction Equipment. April 23, 2003. Vol. 106, Iss. 4.

[42] Stewart, Larry. "Production Heroes: Take the Textbook to the Trench." Construction Equipment. April 2003. Vol. 106, Iss. 4.

[43] We assume a machine using 2 gallons per hour, operating 1,500 hours per year.

Savings for track-type tractors would be well over a thousand dollars per year with such increases in operating efficiency.[44]

With respect to air quality, improved digging methods that reduce the operation time for a backhoe by a single day would reduce emissions of PM by 148 grams, NOx emissions by 1,241 grams, CO emissions by 522 grams, and HC emissions by 148 grams.[45]

2.3.3 How to do it

Contractors can send employees to training programs sponsored by manufacturers, or set up their own in-house training and certification programs. Some companies that experience a slowdown during winter months find that employee downtime can be utilized to enhance skills.

Equipment operators learn a variety of skills during a typical operator training course, such as how equipment can be most efficiently and safely operated. According to Terry Goff, Director Public Policy & Regulatory Affairs at Caterpillar, "Training can help to improve both production techniques and planning techniques. Job planners can learn how to optimize the location of loading tools, and operators can learn how to correctly position hauling tools…For instance, an on-highway truck driver

> *"Training will allow operators to see problems before they happen. The techniques learned will help to reduce fleet maintenance costs. In addition, operators will get more tons per cycle from a wheel loader or will be able to load one more truck an hour with an excavator. These are quantifiable benefits that will increase a company's bottom line and its overall fuel efficiency."* George Schulz, Certified Dealer Instructor, Giles & Ransome, Inc.[46]

pulls exactly to the right place so the wheel loader doesn't have to move more than is necessary…Properly trained operators can save both time and fuel."[47]

In addition, operators can learn to recognize abnormal equipment operation and identify maintenance problems. For dump trucks, trained operators can help maintain the correct tire pressure for the load carried and site soil conditions.

Specific training will vary by equipment type. It could include such operational practices as:

- Pulling trucks along side an excavator where they can be loaded over the tailgate;
- Progressive shifting;
- Digging within a machine's power band;
- Setting the correct work mode, boom priority and swing priority modes for a job; and
- Front to back or slot dozing.

---

[44] Stewart, Larry. Construction Equipment. June 2000. Vol. 101, Iss. 6.
[45] We assume an uncontrolled backhoe loader, with an 89 horsepower engine, operating under a load factor of 0.21. Data obtained from EPA report, Exhaust and Crankcase Emission Factors for Nonroad Engine Modeling— Compression-Ignition. Report No. NR-009c, Revised April 2004.
[46] "Giles & Ransome Touts CAT Certified Operator Training." ConstructionEquipmentGuide.com. 6/4/2003.
[47] Phone conversation, January 19, 2007.

Equipment operating techniques such as progressive shifting (a technique for changing gears) can reduce fuel consumption by allowing a dump truck to be operated in a more fuel efficient gear. Other practices like loading over the tailgate or digging within a machine's power band may optimize production and equipment performance. A bulldozing technique called "slot dozing" requires the operator of the grading equipment to repeat passes in a single blade width before moving over and repeating the process to create an adjacent slot. As each slot deepens, the sides hold material on the blade, and each cut moves more dirt than the last. "A lot of operators who are doing production dozing think they should make long cuts. They want to see dirt boil or roll in front of the blade. In reality, the blade will only hold so much dirt. Once it's full, you start losing dirt off both sides of the blade, leaving windrows that will have to be moved again. You should be able to get a full load on the blade in two lengths of the tractor at the most — and that applies to any size tractor. Once the blade is full, you stop cutting and slide the dirt in front of the blade."[48] Training operators to use techniques like these helps them increase machine productivity and reduce emissions per work task.

Companies can use equipment operational data to measure driver performance and identify operating behavior that can be improved. For instance, companies can track fuel use per hour by operator, and then compare their performance against others operating in similar conditions with the same equipment. Information will help drivers improve their skills to maximize fuel economy. Electronic engine controls can track how often a vehicle is operated in its most fuel efficient mode. The data can be displayed in histograms to help drivers see how they are doing. Drivers who are underperforming can be given additional training. Most new equipment has electronic engine controls. It is easier to track the performance of equipment that is operated in a more routine fashion, such as dump trucks, than equipment with variable engine load and usage patterns, such as earth moving machines. Identifying appropriate metrics to identify and track efficient operation for these types of equipment is more difficult.

Incremental improvements in equipment operations through training can translate into significant improvements in profit and environmental performance.

---

[48] Stewart, Larry. "Doze More Dirt." Construction Equipment. November 1, 2005, http://www.constructionequipment.com/article/CA6280168.html.

3 Fuel Strategies

This section describes two primary fuel strategies to reduce diesel emissions: (1) use of ultra-low sulfur diesel (ULSD), and (2) biodiesel. Overall, the use of ULSD may be the top choice for most companies, because it is widely available and can be implemented relatively easily. Biodiesel can have operational issues that should be addressed.

The table below summarizes the costs and benefits of both fuel strategies. Sections 3.1 and 3.2 provide more detailed information on the fueling options, including their costs, benefits and how to do it.

Fuel Strategies Summary

Fuel Strategy	Costs	Benefits
Ultra-low Sulfur Diesel	Slightly higher price than regular nonroad diesel	Reduces PM emissions Reduces engine wear, corrosion, and deposits May allow increased oil change interval Enables the use of advanced technologies to reduce PM and NOx
Biodiesel (B20, B5)	Slightly higher price than regular nonroad diesel in most regions May increase NOx emissions Small power loss	Reduced PM, CO, and HC emissions May improve lubricity and reduce engine wear

Fuel additives, including emulsified diesel, are not discussed in this report because they do not appear practical for nonroad construction equipment at this time. EPA maintains a fuel registration program that requires manufacturers of all onroad fuels and fuel additives to register their products prior to sale. Information on this program can be accessed at:

http://www.epa.gov/otaq/additive.htm

EPA and the California Air Resources Board (CARB) also have technology verification programs that include fuels and fuel additives. Information on these programs is available at:

http://www.epa.gov/otaq/retrofit/retroverifiedlist.htm

3.1 Ultra-low Sulfur Diesel

ULSD is a refined, cleaner diesel fuel that can be used in any diesel engine. ULSD reduces maintenance costs and harmful emissions and is mandated for onroad use. Vehicles equipped with particulate traps require the use of ULSD, because these emission-control devices can be damaged by sulfur. Diesel oxidation catalysts operate more effectively with ULSD. Nonroad equipment operators currently have the option to select between regular nonroad diesel, low-sulfur diesel (LSD), and ULSD. Regular nonroad diesel has a sulfur content of 3,000 to 5,000 parts per million (ppm). LSD fuel has a sulfur content of 16 to 500 ppm, while ULSD has a sulfur content of just 15 ppm or less. Beginning in June 2007, the use of LSD will be mandated for nonroad engines, and ULSD will be required starting in June 2010.

3.1.1 Costs

U.S. Energy Information Administration data show a $0.05 price difference between ULSD and standard nonroad fuel for October 2006.[49] Some industry analysts project that any price differential between LSD and ULSD will disappear overtime.

3.1.2 Benefits

Using ULSD fuel on its own, without particulate filters or oxidation catalysts, can reduce PM emissions between five and nine percent depending on the baseline fuel sulfur levels.[50] If ULSD is used with particulate filters, PM reductions of 55-90 percent can be achieved.[51] When ULSD is used with oxidation catalysts, PM reductions of 10-50 percent are possible.[52]

ULSD enables the use of advanced emission-control devices in equipment. Using such devices reduces emissions of PM, HC, and precursors of ozone to near-zero levels.[53] Advanced emission control systems required for the 2007 highway engines and future nonroad engines will not operate properly without ULSD. Sulfur poisons the catalytic material on particulate filters and catalysts that are used to burn the particulates to ash. If regeneration doesn't occur sufficiently in PM filters, the filter can become clogged.

ULSD reduces engine wear, deposits and oil degradation. The estimated maintenance savings is more than three cents per gallon compared to regular high sulfur fuel.[54] These savings result from companies' ability to extend oil change intervals. "The reduced sulfur will benefit our EGR [exhaust gas recirculation] and pre-EGR engines by reducing corrosion and extending oil drain intervals" notes Jerry Wang, senior technical advisor for chemistry, fuels, and lubricants at

---

[49] No. 2 Distillate Prices by Sales Type, http://tonto.eia.doe.gov/dnav/pet/pet_pri_dist_dcu_nus_m.htm.

[50] Clean Construction USA, EPA, http://www.epa.gov/cleandiesel/construction/strategies.htm.

[51] Clean Construction USA, EPA, http://www.epa.gov/cleandiesel/construction/strategies.htm.

[52] Clean Construction USA, EPA, http://www.epa.gov/cleandiesel/construction/strategies.htm.

[53] Clean Diesel Fuel Alliance Information Center, "ULSD Issue Paper" available at http://www.clean-diesel.org/images/ULSD_issue_paper.pdf.

[54] This estimate is from the cost-benefit analysis for EPA's rulemaking for "Control of Emissions of Air Pollution from Nonroad Diesel Engines and Fuel" found at 69 FR 38957.

Cummins Engine Company[55] Use of ULSD will enable an oil change interval extension of approximately 35 percent longer than that required for nonroad vehicles using high sulfur fuel.[56] The maintenance savings can help offset the higher price paid for ULSD.

3.1.3 How to do it

ULSD is stored in the same tanks and uses the same fueling systems previously used for regular diesel fuel. Currently, ULSD is available everywhere for highway use.

When switching to ULSD some fleets have changed fuel filters after two or three tanks of fuel because they have been concerned with the cleaner fuel acting as a solvent to remove sediment from fuel tanks. EPA has not been able to document this as a common occurrence; nevertheless, it may be a practice to consider.[58] Terry Goff, Director of Public Policy & Regulatory Affairs at Caterpillar notes, "Caterpillar is aware of many customers who have used ULSD in nonroad equipment without problems. We recommend that users consult the fluid guide produced by their equipment manufacturer. The fluid guide addresses equipment specific information such as ULSD's impact on O-ring seals." [59]

> "We've been using ULSD since July 2001 and we haven't had any trouble with it….It's been transparent to us as users – ULSD ends up performing just like our old diesel. And that's all we care about." David Kerrigan, fleet services director for the City of Seattle[57]

In the past, lubricity was a potential concern associated with ULSD's performance Lubricity is a measure of the fuel's ability to lubricate and protect the various parts of the engine's fuel injection system from wear. Low lubricity can potentially increase wear on certain fuel injectors. Sulfur in fuel acted as a lubricity agent. The fuel processing required to reduce sulfur to 15 ppm removes naturally-occurring lubricity agents in diesel fuel. In order to address this concern, the American Society for Testing and Materials (ASTM) has adopted the lubricity specification defined in ASTM D975 for all diesel fuels. The standard went into effect January 1, 2005. Fuels that meet this specification should have the required lubricity for proper engine operation. Any fuel lubricity additives necessary to meet the new specification are being added by the fuel suppliers, so end-users do not need to add fuel lubricity additives.[60]

---

[55] Kilcan, Sean. "Ultra Low Impact." Fleet Owner. June 2003.Vol.98, Iss. 6.
[56] Id.
[57] Kilcan, Sean. "Ultra Low Impact." Fleet Owner. June 2003.Vol.98, Iss. 6.
[58] http://www.epa.gov/cleandiesel/maintenance.htm.
[59] Phone conversation, January 19, 2007.
[60] Additional information on the introduction of ULSD is available from the Clean Diesel Alliance at the following address: http://www.clean-diesel.org/.

3.2 Biodiesel

Biodiesel is a renewable fuel made from domestically grown crops such as soybeans, cottonseed, peanuts, and canola, as well as other biotic materials, including recycled cooking grease. Biodiesel is usually blended with petroleum diesel. B5 and B20 are common blends. B5 is a blend of five percent biodiesel and 95 percent petroleum diesel, while B20 contains 20 percent biodiesel and 80 percent petroleum diesel.

3.2.1 Costs

The price of biodiesel depends on the production process used, the distribution and blending costs, and the feedstock employed. Prices vary among regions. According to the most recent Clean Cities Alternative Fuel Price Report from February 2006, the national average price for B20 was $2.64, while the average price of diesel was $2.56. On average, biodiesel costs approximately $0.08 more per gallon. In the Midwest, average biodiesel prices were within $0.02 of regular diesel. In some regions biodiesel may be less expensive than ULSD.[61] Biodiesel is often subsidized by governments to encourage its use.

The use of biodiesel may increase NOx emissions. EPA is currently conducting an analysis of the potential NOx increase with biodiesel. Specific NOx increases depend on the fuel blend used, equipment type, and operating patterns of the equipment/vehicle. On average, past data indicates that the use of B20 resulted in a NOx emissions increase of about two percent.[62] Biodiesel contains less energy than petroleum diesel. Power, torque, and fuel economy differences are between one and two percent with B20, which are not noticeable for most users.[63]

3.2.2 Benefits

Biodiesel may provide lubricity and several other advantages. Some organizations using biodiesel claim to have experienced savings due to a reduced number of fuel system failures, as well reductions in other types of maintenance problems. Using biodiesel may reduce carbon monoxide, hydrocarbon and particulate matter emissions. Biodiesel may have a cleaning effect on the engine, resulting in an engine that produces less smoke, runs smoother and with less noise.

B20 may reduce PM emissions by up to 10 percent, in addition to HC and CO emissions.[64] Specific emissions reductions are dependent on the feedstock used to make biodiesel and other factors. B20 can also reduce life cycle CO_2 emissions, since its production employs a closed carbon cycle that grows and processes plants to produce new fuel.[65] EPA has a calculator on its

---

[61] Clean Cities Alternative Fuel Price Report, February 2006, http://www.eere.energy.gov/afdc/resources/pricereport/pdfs/afpr_feb_06.pdf.
[62] EPA. A Comprehensive Analysis of Biodiesel Impacts on Exhaust Emissions. EPA420-P-020001, October 2002.
[63] U.S Department of Energy, 2004 Biodiesel Handling and Use Guidelines. DOE/GO-102006-1999, http://www.nrel.gov/docs/fy06osti/39451.pdf.
[64] http://www.epa.gov/otaq/retrofit/techlist-biodiesel.htm.
[65] EPA. A Comprehensive Analysis of Biodiesel Impacts on Exhaust Emissions. EPA420-P-020001, October 2002.

web site that helps companies calculate the emissions reductions from the use of different blends of biodiesel in different types of equipment. It can be accessed at:

http://www.epa.gov/otaq/retrofit/techlist-biodiesel.htm

3.2.3 How to do it

Companies that use biodiesel should use fuel grade biodiesel that meets the ASTM D6751 standard. Users of non-fuel grade biodiesel have reported a number of problems with fuel quality.

Most nonroad vehicles can run on B5 with little modification. B20 is also commonly used in larger equipment. Pure biodiesel can soften and dissolve some rubber, including vehicle fuel lines and pump seals. On older vehicles, it may be necessary to replace the fuel lines and other components. Biodiesel can also clean injectors and fuel lines. The first few tanks of biodiesel may loosen accumulated deposits and clog the fuel filter. As a result, users may want to consider replacing the fuel

> "In the construction business, green has become a priority. Our biodiesel strategy has helped position our company at the top." Tom Ambrey, CEO, RAFN Construction Bellevue, WA[66]

filters after the first couple of tanks of biodiesel. There have been many reports of engines having fuel injector pump failures when using biodiesel due to O-ring shrinkage. This may be caused by the lower aromatic content of biodiesel compared to normal diesel, and the fact that nitryl rubber O-rings absorb aromatics. This problem has been solved in newer engines that use Vitron O-rings, which are much less susceptible. Most manufacturers suggest that engines should be monitored during any fuel switch by using oil sample analysis and watching for leaks or performance problems.

Pure biodiesel has a higher gel point. B100 (100 percent biodiesel) starts to become viscous at 32°F. For these reasons, biodiesel is usually blended with petroleum diesel. B20 has a gel point of -15°F. In order to ensure winter operability, some users in very cold environments store vehicles in or near buildings, use fuel heaters, and employ cold flow additives.

Long term storage of biodiesel in tanks is more susceptible to the formation of bacteria and algae. It is recommended that periodic testing be done to ensure microorganisms are not present in biodiesel storage tanks. Tanks can be treated with biocides to prevent bacteria growth.

Biodiesel is available from distributors and retailers in most states. A list of biodiesel retailers and distributors can be found on the Internet at the sites shown below.

http://www.biodiesel.org/buyingbiodiesel/retailfuelingsites/default.shtm

http://www.biodiesel.org/buyingbiodiesel/distributors/default.shtm

---

[66] http://www.propelbiofuels.com/site/.

Most major engine companies have stated formally that the use of blends up to B20 will not void their parts and workmanship warranty. Recommended practices for use of biodiesel may vary by model. For example, Caterpillar recommends using a maximum of a B5 blend for engines up to 150 horsepower and up to a B20 blend in engines over 150 horsepower. Case has approved the use of biodiesel blends up to five percent in all their mechanical engines.[67] Grant Goodman, Owner of Rockland Materials notes, "Rockland Materials has been operating loaders, excavators, rock-crushers, mining equipment, and generators (gen-sets) on biodiesel…without incident as long as it's been operating trucks on the fuel. Using biodiesel has not affected equipment warranties, either….We have the blessings from Caterpillar and Cummins to do that."[68] It is important to check with the manufacturer before using biodiesel. Formal statements from manufacturers have been compiled and are available on the National Biodiesel Board's web site at the following address.

http://www.biodiesel.org/resources/fuelfactsheets/standards_and_warranties.shtm

---

[67] "Case Approves Biodiesel". Construction Equipment. Vol. 109, September 2006.
[68] "Biodiesel: A Cleaner, Greener Fuel for the 21st Century." Environmental Building News. January 2003.
http://www.oregon.gov/ENERGY/TRANS/docs/biodiesel.pdf.

4 Equipment Strategies

This section describes three equipment modification strategies for reducing diesel emissions: (1) installing retrofit (i.e., exhaust emission control) devices, (2) repowering or upgrading older diesel engines, and (3) using grid electricity or hybrid electric equipment. Although these equipment strategies often require a significant initial investment, they can be extremely cost-effective methods for improving air quality. In some cases, public funds may be available to subsidize the cost of retrofits or repowers. In November 2006 the Diesel Technology Forum (DTF) released its report, *Retrofitting America's Diesel Engines - A Guide To Cleaner Air Through Cleaner Diesel*, which provides an extensive list of federal and state incentive and funding resources.[69] Construction companies are positioning themselves to win new business by voluntarily reducing emissions from their nonroad fleet.

Repowering construction equipment improves performance and longevity, but it is costly. Some companies have found emissions upgrade kits to be a lower-cost way to improve the emissions performance of equipment that is destined to be rebuilt. Another strategy is using electric grid power. Electrification can save companies money and lower emissions per kilowatt hour, when it is a viable option.

The table below summarizes the costs and benefits of each equipment strategy. Sections 4.1 through 4.3 provide more detailed information on each equipment option, including their costs, benefits and how to do it.

Equipment Strategies Summary

Equipment Strategy	Costs	Benefits
Retrofit (Exhaust Emission Control) Technologies	Retrofit technology and installation costs, which can be subsidized with public grant money in many cases Some retrofit equipment should be used with ULSD, which may have marginally higher cost	Reduced PM, NOx, CO, and HC emissions Positioned to win new business on contracts requiring cleaner construction equipment
Engine Repower or Upgrades	Cost of replacing an engine, which can be offset with public grants in some cases Engine emissions upgrade kits cost less than replacing an engine	Reduced PM, NOx, CO, and HC emissions Lower fuel consumption Improved engine reliability and lower maintenance costs

---

[69] DTF's full report is online at: http://www.dieselforum.org/newsarticle/article/641/1/.

		Positioned to win new business on contracts requiring cleaner construction equipment
Electrification	Costs of using grid power when it is available Purchase of electric or hybrid electric equipment	Reduced PM, NOx, CO, and HC emissions Grid power has lower per kilowatt-hour cost Hybrid electric vehicles have substantially lower fuel consumption

4.1 Retrofit Technologies

Diesel retrofit technologies are devices that are attached to equipment to remove pollutants from engine exhaust. The two most common diesel retrofit technologies are diesel oxidation catalysts (DOCs) and diesel particulate filters (DPFs). DOCs and DPFs reduce PM, CO, and HC emissions. DOCs can be used not only with conventional diesel fuel but have also been shown to be effective with biodiesel and other alternative diesel fuels. DPFs remove particulate matter in diesel exhaust by filtering exhaust from the engine. DPFs require the use of ULSD fuel.

4.1.1 Costs

Oxidation catalysts currently cost between $1,000 and $2,000 and can be installed on almost any new or used engine.[70] Diesel particulate filters currently cost between $5,000 and $10,000 and can generally be installed on certain vehicles with engines built after 1995.[71] To ensure a DPF will work properly, it is necessary to use ULSD fuel with a sulfur content of less than 15 parts per million.

The installation of the retrofit device is an additional cost. Installation fees are typically higher for nonroad equipment than for highway trucks, because there is a larger array of equipment types and operating conditions. The costs associated with equipment downtime during installation can be significant and vary by equipment type and level of utilization.

> "We have demonstrated diesel particulate filters on different engines, different equipment, a variety of operating environments…the equipment is more than capable of handling anything that we need." Ramesh Raman, Project Coordinator, MTA Capital Construction[72]

While these technologies are not inexpensive, there are a number of programs available to help companies pay for them. Where public grant money is available, companies can apply to get most or all of the cost of the equipment and its installation paid for with public assistance. For example, grants to install retrofits can be obtained from the California Carl Moyer Memorial Air Quality Standards Attainment Program (the California Carl Moyer program) and the Texas Emissions Reduction Program (TERP), as well as from other states such as New Jersey, Oregon and Tennessee. Federal money is also available through the National Clean Diesel Campaign and the Congestion Mitigation and Air Quality (CMAQ) program administered by U.S. Department of Transportation. EPA provides a list of funding sources at the following address:

http://www.epa.gov/cleandiesel/construction/grants.htm

---

[70] EPA Region 1, http://www.epa.gov/ne/eco/diesel/retrofits.html#edf.
[71] EPA Region 1, http://www.epa.gov/ne/eco/diesel/retrofits.html#edf.
[72] Phone conversation, Ramesh Raman, Project Coordinator, MTA Capital Construction, January 18, 2007.

4.1.2 Benefits

Investing in these equipment upgrades may position a construction company to bid on construction contracts that call for the use of cleaner equipment. For example, New York City has a new law requiring clean fuels and retrofits on city contracts. Several public agencies – Connecticut Department of Transportation, Massachusetts Highway Department, and Massachusetts Bay Transportation Authority – now require low emission equipment on construction projects. In the Midwest, the $6.6-billion O'Hare Modernization Program in Chicago is requiring use of ULSD fuel and DOCs on mobile equipment. In certain cases, areas that do not meet air quality standards are willing to provide companies with monetary incentives to retrofit their equipment. This provides an opportunity for contractors to modernize the emission technology on their entire fleet.

Diesel oxidation catalysts can reduce emissions of PM between 20 and 40 percent, HC by 50 percent, and CO by approximately 40 percent from current levels for a typical piece of equipment. Diesel particulate filters physically trap and oxidize PM as exhaust gas flows through them. DPFs reduce PM emissions by approximately 90 percent, HC emissions by 60-90 percent, and CO emissions between 60-90 percent.[74]

> *"This is the way of the future…with all the air-quality standards coming out limiting smoke, there are jobs you won't be able to get if you don't have clean engines."* Mike Bowman, Equipment Manager, Coburn Equipment[73]

There are several examples of construction projects using retrofitted equipment. The Central Artery/Tunnel Project in Boston used more than 200 pieces of construction equipment retrofitted with DOCs. Estimates for the years 2000-2004 show emission reductions of 36 tons/year of CO, 12 tons/year of HC, and 3 tons/year of PM. Other projects using retrofitted construction equipment include the Dan Ryan Expressway in Illinois, the I-95 New Haven Harbor Crossing Corridor Improvement Program in Connecticut, the 7 World Trade Center site and MTA Capital Construction projects in New York.[75]

4.1.3 How to do it

EPA's National Clean Diesel Campaign and CARB have technology verification programs that evaluate the emission reduction performance of retrofit technologies, including their durability, and identify conditions that must exist for these technologies to achieve those reductions.[76] CARB ranks devices by their level of effectiveness in reducing emissions. EPA assigns reduction values based on the test results during verification.

---

[73] Stewart, Larry. "Clean-Engine Replacements Hone Fleet's Competitive Edge." Construction Equipment. May 1, 2003. ConstructionEquipment.com.
[74] Retrofitting America's Diesel Engines: A Guide to Cleaner Air Through Cleaner Diesel. November 2006. Diesel Technology Forum, http://www.dieselforum.org/fileadmin/templates/Resources/RetrofitMaterialsFactSheets/Retrofitting_America_s_Diesel_Engines_11-2006.pdf.
[75] MECA. Case Studies of Construction Equipment Diesel Retrofit Projects. March 2006, http://www.meca.org/galleries/default-file/Construction%20Case%20Studies%200306.pdf.
[76] For a list of verified technologies, see: http://www.epa.gov/otaq/retrofit/retroverifiedlist.htm.

Retrofit technologies must fit the equipment application. Some technologies have exhaust temperature requirements to enable them to achieve the greatest emissions reductions. Suppliers of retrofit technologies should evaluate the exhaust temperature profiles of the targeted fleet to determine which technologies perform appropriately. For example, passive diesel particulate filters need to be operated above a certain temperature to ensure regeneration. Suppliers need to consider this when retrofitting nonroad equipment that experiences frequent periods of low-load operation or idling, when exhaust temperatures may drop to 150°C or lower for a considerable length of time. Data logging of exhaust temperatures is usually necessary to determine the suitability of diesel particulate filters.

Selection of the most appropriate technology depends on the type of equipment being retrofitted and the operating conditions. Installations need to take into account severe environmental conditions, operator visibility, space and weight constraints, and placement issues. Retrofits are becoming more common among nonroad fleets, and as a result, suppliers are becoming skilled at addressing the challenges of retrofitting nonroad equipment.

EPA's Retrofit Technology Verification Program groups together engine families with similar emissions performance characteristics, making it easy to select retrofit technologies available for construction equipment. EPA maintains a list of verified retrofit technologies available at:

http://www.epa.gov/otaq/retrofit/retroverifiedlist.htm

EPA also recognizes and accepts those retrofit devices that have been verified by CARB. Information about CARB's Verification Program and links to their list of verified technologies can be found at:

www.arb.ca.gov/diesel/verdev/home/home.htm

A list of retrofit manufacturer contacts is available at:

http://www.epa.gov/otaq/retrofit/cont_retromfrs.htm

4.2 Engine Repower or Upgrades

Engine replacement, or "repowering," is the replacement of an older diesel engine with a new, lower emission engine system. "Upgrading" means adding emissions-reducing parts, most often during an engine rebuild. Repowering older construction machinery is becoming a fairly common practice in states such as California and Texas. It can be cost effective and economical in some cases. Upgrading an engine at the time of rebuild allows companies to modernize equipment for fairly low marginal cost. Engine manufacturers may offer an emissions upgrade kit that allows dealers to overhaul an engine with new components to improve emission performance and extend the engine's useful life.

4.2.1 Costs

The cost effectiveness of repowering a piece of equipment depends on the make and model of the machine and the availability of grant money to defray these costs. "Voluntary retrofit programs are flourishing nationwide, enticing construction companies to apply for grants. For Sukut Equipment Inc, the cost to repower single-engine scrapers tops out at $120,000. It receives money from a state agency that kicks in $96,000. They are picking up almost two thirds of the cost of repowering a machine", notes Mike Ortiz of Sukut Equipment Inc.[78] The repower

> "Cleaner engines can mean improved fuel economy and reduced fuel costs. Participation also signals to the local community a commitment to environmental improvement."
> Terry Goff, Director Public Policy & Regulatory Affairs, Caterpillar [77]

costs need to be compared to the cost of buying new equipment. Smaller equipment would theoretically have lower repowering cost. Installing a new engine in a typical D6H track-type tractor costs about $27,000.[79] These estimates include both the cost of the engine and installation costs.

Emissions upgrade kits can be a cost effective way for companies to upgrade equipment while rebuilding an engine. Caterpillar sells an engine upgrade kit that provides a low cost way to upgrade emissions controls during an engine overhaul. These "emissions upgrade groups" include an upgraded turbocharger, fuel pumps/governor, nozzles, cylinder packs, and installation parts. Installing an emission upgrade kit during the engine rebuild for a typical track-type tractor could add several thousand dollars to the cost of the engine rebuild, but reduce PM, NOx, CO and HC.[80]

---

[77]Conversation with Terry Goff, Caterpillar, Director Public Policy & Regulatory Affairs, Power Systems, Caterpillar Inc.

[78] Independent Construction Caterpillar 633D Scraper Tier 2 Engine Repower Fact Sheet. http://www.airquality.org/ceqa/ProjectsfundedwithMitFees.pdf.

[79] Conversation with a Caterpillar Dealer, January 18, 2007.

[80] Conversation with a Caterpillar Dealer, January 18, 2007.

4.2.2 Benefits

Engine replacements and upgrades can help construction companies win new business and modernize their fleets. Many agencies across the U.S. are writing air quality requirements into public construction contracts. Some contractors are moving to upgrade the emissions control technology in their fleets now while grants or other incentives are available to defray the costs. They believe it will help them stay ahead of the curve if new emissions regulations mandate this technology in the future.[81] Mike Crawford, CEO of Sukut Equipment Inc. notes, "We are using as much capital as we can afford to upgrade our engines....That is going to be the price of admission to stay in this business."[82]

Replacing an engine can extend equipment life, improve fuel economy and lower maintenance costs, thus reducing overall equipment operating costs. "We're not only reducing our emissions, but we're going to have fewer problems with our fleet. We're taking eight tractors and making them like new again", says Mike Bowman of Coburn Equipment[84]

> *"In prior years, we were offered a job or two per year on the kind of project that requires lower machine emissions....Since we started letting people know three months ago that we would have repowered tractors, we've gotten three offers to work on projects where the machines have to run clean."* Sandi Capel, General Manager of Coburn Equipment[83]

For instance, manufacturers have reported that new engines have shown fuel economy improvements of five percent or more.[85] A Track type tractor using 16 gallons of fuel per hour, and operating 1,500 hours per year, could reduce fuel costs by $3,000 per year by repowering with an engine that is five percent more efficient.[86] Cost savings for repowering older equipment that is less fuel efficient would be significantly greater.

Emissions reduction benefits of engine replacement depend on the originally certified emissions level of the vehicle and the replacement engine. Average emissions reductions vary from 25 percent up to 75 percent.[87]

Upgrading the emission control system during an engine rebuild can provide significant benefits to contractors. By spending a few thousand dollars more during an engine rebuild, owners are

---

[81] Hampton, Tudor. "Regulators Want Old Diesel Engines to Clean Up....Or Clear Out." ENR. May 29, 2006. Vol. 256, Iss. 21., http://enr.ecnext.com/free-scripts/comsite2.pl?page=enr_document&article=feenar060529.

[82] Hampton, Tudor. "Regulators Want Old Diesel Engines to Clean Up....Or Clear Out." ENR. May 29, 2006. Vol. 256, Iss. 21.

[83] Stewart, Larry. "Clean-Engine Replacements Hone Fleet's Competitive Edge." Construction Equipment. May 1, 2003. ConstructionEquipment.com.

[84] Stewart, Larry. "Clean-Engine Replacements Hone Fleet's Competitive Edge."Construction Equipment. May 1, 2003. ConstructionEquipment.com.

[85] Power Source. John Deere. Vol 4, 2005.

[86] Diesel costs are assumed to be $2.50 per gallon. Fuel usage varies widely by equipment size, type and usage. The fuel economy benefits for larger pieces of equipment or equipment that is used more intensively could be significantly greater.

[87] Cleaner Diesel Handbook. April 2005.
http://www.environmentaldefense.org/documents/3992_DieselHandbook_CostEffectiveness.pdf#search=%22Cost%20to%20repower%20construction%20equipment%22.

able to ensure that rebuilt equipment can be used on contracts requiring clean equipment.[88] Having equipment that meets more stringent air quality standards can increase resale value of the equipment. Some manufacturers provide emissions upgrade kits that can be used during an engine rebuild. Caterpillar's emission upgrade kit has been verified by the EPA to reduce PM emissions by 15 percent, CO by 3 percent, HC by 61 percent, and NOx by 27 percent.[89]

4.2.3 How to do it

There are a number of factors to consider when repowering. Owners may need to take the following steps:

- Consult with the equipment and engine manufacturers to select a repower engine arrangement to match power and torque curves for application;
- Address installation issues and re-engineering needs;
- Maintain inlet manifold temperature and other parameters per the repower engine's original certification; and
- Modify the software and electrical system of the older vehicle to adapt to the newer engine.

In some areas grant programs are available to subsidize the repowering of equipment. California's Carl Moyer program and Texas' TERP provide incentive funds for the incremental cost of upgrading to cleaner engines or purchasing and installing retrofit technologies. Other states such as New Jersey, Oregon and Tennessee have recently established programs to encourage retrofits, rebuilds and replacement of nonroad engines. Additional funds may be available through the Diesel Emission Reduction Program of the Energy Policy Act of 2005 and the Congestion Mitigation and Air Quality (CMAQ) program administered by U.S. Department of Transportation. For more information, see:

http://www.epa.gov/cleandiesel/construction/grants.htm

---

[88] Barnes, Jim. "New Life for Old Machines." On-Site. Jan/Feb 2005.Vol.49, Iss. 1. p. 15., http://www.on-sitemag.sartech.ca/.
[89] http://www.epa.gov/otaq/retrofit/techlist-cat.htm#eu.

4.3 Electrification

Electrification involves employing electric or hybrid electric equipment. This section also discusses the use of fuel cells to generate clean electric power at construction sites. These strategies can reduce emissions.

4.3.1 Costs

Using electric grid power instead of on-site diesel generators can be a low cost solution if the electric power can be made available safely.

Currently there is little information on the costs associated with heavy duty electric and hybrid electric vehicles, since many companies are currently in the process of developing these vehicles. Volvo believes it will begin commercial production of hybrid equipment in 2009.[90]

Companies employing fuel cells to generate power on-site have paid $5,000 – $7,000 for fuel cell generators. With hydrogen fuel canisters costing $1,000, these generator technologies may currently only be appropriate for niche applications. Prices will likely come down as the market matures. The Energy Tax Incentives Act of 2005 provides tax credits for the purchase of fuel cell powered generators.

4.3.2 Benefits

Switching from diesel power to electric power nearly always brings reduced on-site emissions and a quieter work site. It may also save money.

Grid power produces fewer emissions per kilowatt hour than power generated with on-site diesel generators.[91] An uncontrolled 60 kilowatt generator operating at 40 percent load produces 73 grams of CO, 337 grams of NOx, and 24 grams of PM per hour. If grid power can be accessed onsite and used instead, CO emissions per kilowatt hour can be cut by 91 percent, NOx emissions by 75 percent, and PM emissions by 98 percent.[92]

Utilizing grid power can also yield significant cost savings on a per kilowatt hour basis. For instance, at full load a 54.6 KW diesel generator can use 4.5 gallons per hour of diesel fuel.[93] Assuming fuel costs approximately $2.50 per gallon, the per kilowatt hour cost of diesel generated power would be $0.205. Grid power is substantially cheaper. At the U.S. average retail

---

[90] "Volvo Group presents hybrid technology in the US". January 11, 2007
http://www.volvo.com/financialservices/netherlands/nl-nl/news+and+media/press+releases/NewsItem.htm?channelId=2052&ItemID=15696&sl=en-gb.
[91] Goldstein, Glenn. Assessing the Regional Implications of Advanced Truck Stop Electrification: A Report to the EPA. February 2003.
[92] EPA. "Emissions Factors for Uncontrolled Gasoline and Diesel Industrial Engines."
http://www.epa.gov/ttn/chief/ap42/ch03/final/c03s03.pdf#search=%22Emissions%20Factors%20for%20Uncontrolled%20Gasoline%20and%20Diesel%20Industrial%20Engines%22.
[93] Diesel Generator Set, Caterpillar
http://www.cat.com/cda/components/securedFile/displaySecuredFileServletJSP?fileId=263952&languageId=7.

price of $0.108 per kilowatt hour, a construction site using a 60 kilowatt load of power for 40 hours a week would save $231 per week by using grid power.[94]

There may be other opportunities for contractors to use electric equipment instead of diesel. For instance, some contractors have found that with stationary equipment, such as power washers, an electric unit has fewer components to maintain and repair. Potential savings and emissions benefits vary by equipment type.

Because the use of hybrid vehicles are in the early stages of introduction, only limited information is available. Fuel efficiency increases of 35 percent or more are predicted by manufacturers.[95]

4.3.3 How to do it

The procedure for electrification depends on the construction process. On construction sites, power is often not available or is turned off for safety reasons. Companies can implement a policy to use grid power whenever it becomes available. In some cases, it may be possible to sequence job activities so that power can be made available earlier in the job.

The use of electric or hybrid-electric technologies in nonroad applications offers great promise. "Electrification will enable our products to have significantly reduced emissions and noise levels with improved efficiency and sustained high performance" notes Bruce Wood, Director of ePower Technologies Group, from John Deere & Company[96] Following are several examples of uses of electric equipment.

- The U.S. construction industry relies heavily on diesel-powered mobile cranes. In Europe, mostly electric-powered tower cranes are used. There may be opportunities for greater use of electric cranes on U.S construction sites.

- John Deere & Company is undertaking several electrification initiatives, including the use of advanced battery technologies, diesel hybrid electric technologies, and hydrogen fuel cell technologies to add performance and efficiency capabilities to existing products. Deere currently has available an electric-powered E-Gator Utility Vehicle. Initial prototypes and demonstrators are in the smaller turf care products, but these technologies eventually will be applied in larger equipment as well.

- Volvo Group has developed a hybrid power system concept that provides fuel savings in transport operations. The engine system is appropriate for commercial heavy construction vehicles. Volvo estimates improvements in fuel economy of 35 percent can be achieved.

---

[94] This is the U.S average retail electricity price as of June 2006. EIA web site: http://www.eia.doe.gov/cneaf/electricity/epm/table5_6_a.html.
[95] Van Hampton, Tudor. "Eco-Friendly Engine Pioneers Are Searching for New Sources of Clean, Green Power" ENR. Feb 12, 2007.Vol.258, Iss. 6; pg. 39.
[96] "John Deer to Produce Fuel Cell Demonstrator Vehicle." OPEESA Industry News. December 20, 2002.

The engines will have lower maintenance costs due to reduced wear and tear on brakes. In addition, hybrid engines will produce less noise.[97]

- John Deere is working with Hydrogenics Corporation, a Canadian fuel cell manufacturer, to develop a technology demonstrator fuel cell-powered Commercial Work Vehicle (CWV). The demonstration model will be a modified John Deere Pro-Gator$^{(TM)}$ Utility Vehicle. The timeline for widespread introduction of this technology is uncertain.

There appear to be some applications for the use of fuel cell generators on construction sites. Providing power for tools or charging batteries used in cordless power tools is often done using standard portable generators. Conventional gasoline or diesel generators cannot be used indoors or on scaffolding. Voller Energy sells a portable fuel cell generator that uses hydrogen to power a battery charger that can be used anywhere, including indoors. The system emits only water. The Coleman Powermate Airgen is another fuel cell generator that is available. The system provides 1kW of uninterruptible power and was designed for sale to industrial customers. The current market only supports niche applications of these products. A number of companies are working on more powerful generators.

---

[97] "Volvo develops hybrid." Pit and Quarry. April 2006. Vol. 98, Iss 10.

5   Conclusions

There are many opportunities for small, medium, and large construction contractors to reduce diesel emissions while respecting their bottom lines. These strategies help achieve important public health and environmental objectives. Many of them also provide additional business benefits such as improving efficiency and lowering operating costs.

Implementing clean construction practices can help companies become good stewards of the environment while maintaining a competitive edge. Because more public contracts are now requiring companies to use retrofits and clean fuels, voluntarily adopting these strategies in advance could position a company to win new business. As the entire construction industry moves toward cleaner operations, these practices can help companies to stay competitive.

The focus of this research has been to describe low cost diesel emissions reduction practices in three primary areas: operational strategies, clean fuels, and equipment strategies. Not every strategy will make sense for every company. Most companies can adopt a number of these practices and achieve both efficiency improvements and emissions reductions.